I
Herring Gull

Despised, hated, abused, starving and culled by humans

MIKE PEARCE

DEDICATION

Dedicated to all those who just think seagulls are vermin. Also for those who can appreciate the needs of the gulls and how humans have influenced the expectations of gulls to receive food from humans. It is especially dedicated to those famous people who have been literally shot down for trying to defend the herring gull as a worthwhile member of our world.

An old Scottish rhyme predicting the weather

Sea gull, sea gull sit on the sand

It's never good weather when you're on the land

From a song - The Seagull by Arthur Askey. 1940

Happy at the seaside never having words

Happy in Bond Street with the other birds

Just a silly seagull, that's all

Fly away Peter, fly away Paul

When we see somebody we like

We're as sweet as treacle

When we see somebody we hate

Plop-goes the seagull

From the poem, The Sea-gull, by Mary Howitt.

The wild sea-gull, the bold sea-gull,

As he screams in his wheeling flight,

As he sits on the waves in storm or calm,

All cometh to him alright!

Nor any his will gainsay!

And he rides on the waves like a bold young king

That was crowned but yesterday.

CONTENTS

ACKNOWLEDGMENTS

The author would like to thank Christine Pearce for
reading and checking the manuscript

PREFACE

Seaside wouldn't be seaside without herring gulls. It would be a desolate place probably only with the noise from a few crows pecking around on the beach. Humans, just as they associate ice-cream, amusement arcades, buckets and spades and lilos etc. with the vista, regard seagulls as a trigger of past memories.

As a seagull, you can live for over forty years old but probably less in an urban environment. This book links you with the life of a herring gull looking at how you behave and the challenges that you must face in your daily life. Hopefully it will help change your view that we are not just pesky birds all out to cause damage and havoc amongst tourists.

1 INTRODUCTION

Oceans in the past provided us with an abundance of food. This has all changed with the depletion of fish stocks. We used to depend a lot on man's activities involved with food and shelter. Man's fishing activities provided us with security in finding food at the quayside and fish processing areas. This was supplemented by huge landfill sites and rubbish dumps as well as by the ploughing of acres of land given to crops.

We had plenty of food to feed on, there being little direct contact between ourselves and humans. With recycling and the reduction of farming activities, the land being used for houses motorways etc., we have been left in some parts of the country in a famine situation finding it difficult to feed our youngsters. Our mournful cry once presented a happy call to others but now can be seen as a cry for help. Under the present climate we must steal and beg. We are being persecuted as vermin and labelled slaughtering seagulls, sinister, a big evil, viscious flying rats, violent creatures, screeching banshees, neighbours from hell, winged maniacs, demons of good old blighty, savage beasts, and sadistic blood thirsty jarring screechers by many who only see us as a threat.

2 WE HAVE BEEN AROUND MUCH LONGER THAN YOU

We are very adaptable and bold. No one can say we have not survived successfully for thousands of years. We evolved over 15 million years ago in the North Atlantic, nesting on the cliff faces around the coast. Our ancestors were slow movers compared to us and many were plant eaters being derived from dinosaurs, as Aldous Huxley suggested. With our large wings, not membranous like pterodactyls, we could escape the threats from predators or changes in the environments. Our feathers helped us to survive wintry conditions and we did not need to warm up using the heat from the sun like our reptilian ancestors. We could migrate for miles away from creeping ice sheets to different areas to obtain food and bring up our young.

There are still around 2500,000 in the world today but numbers are decreasing. We belong to this world as much as any other animal does who has striven to exist over this long time. We have always been great opportunists, especially where food is located and being consumed by other animals. There is always extra and waste. We have always been able to exploit this especially when humans came on the scene. The origin of our name may be linked to the old Norse word for herring meaning army but many link us with the times of herring abundance in the sea. In Irish Celtic mythology we have been considered a god of the sea, others see us as spiritual messengers but this is just imagination.

Although I say it, we are beautiful birds and keep our feathers clean. Our feathers are as white as new-fallen snow. You show me a dirty seagull! We can reach over a thousand grams in weight. Our beaks are bright yellow with a red spot. We even have yellow skin around our eyes. Our wingspan can reach as much as five feet. We love to sail in the thermals and perform acrobatics to dive down and pick things up. The wind generated by waves can also provide us with lift. Many say we are clever, crafty and brazen. Unfortunately we have a problem with our feet. They are ideal for paddling along in the sea and balancing on flat surfaces but not much good for grasping and holding, especially on buildings and city furniture. We do, though, have good weight distribution which allows us to perch on top of thin objects and the heads of statues so as to get an all-round view. Many a statue of a famous person has one of us sitting or standing proudly on top.

3 COMING TO LIVE IN A PLACE NEAR YOU

Satellite groups of us exist along the coast. We have been named coastal scavengers. We feed and rest floating on calm seas or sitting on turf covered cliffs or roofs. In rough weather, we can always return to land. We love wide, open spaces such as parks, car parks or playing fields where we can see any danger or threats in the distance. You can find us often at quays and harbours, especially where fish are being loaded and unloaded.

We prefer rocky outcrops and cliffs but also stay around lakes and some rural habitats. We have though modified our behaviour for an urban environment where many of us are permanent residents now. Where our populations have increased, then our young can move to other areas to colonise where food is abundant. Tall buildings are cliff top substitutes which help our young survive and avoid foxes and cats. We used to migrate but have become used to urban living where we were born. Roofs, parapets and chimneys offer us some shelter and warmth which is better than nesting on a cliff top where we are exposed to gales and harsh rain. We, love flat and corrugated roofs to loaf around on, especially if you are a male like me and we can here search for food amongst the moss and lichens. You may have noticed lumps of these thrown off roofs onto the ground. That's us poking around. Some of the new buildings, however, do not allow us access, especially where we need to flap our wings to take off. Many buildings in cities are very high and impossible to nest on but they do give good up - draughts which we can use for soaring over. Our rural relatives sometimes come inland to see us and breed but we don't mix. We are not very social like some other animals.

European gulls migrate further south in winter but many of us are permanent in many countries including Britain.

In winter or bad weather we group together on beaches and open areas withstanding all kinds of weather. Many of us may be injured or unwell from eating contaminants containing bacteria or poisons such as lead

4 CALLING, TALKING AND NESTING

You can hear many of our different calls especially when we are in large groups and squabbling. We make a lot of noise especially at landfill sites but also are noisy when we know it is about to rain.

Some say we have over ten different calls. We have a food finding call but only when the food is in large supply. This also attracts others, often our relatives. We feel safer if other gulls are with us, especially when feeding. If it is a small amount of food then we often won't call. We gulls don't share food even with fledglings

If we see a threat one of us will emit a bark like an alarm call often consisting of two descending notes then three staccato single notes if there is extreme danger. We use this especially when our chicks are around and others will come to our aid and start barking as well. The alarm call can also be used to cause thousands of seagulls to leave an area.

Humans say we make a real din in breeding areas with our laugh like trumpeting which causes others to copy and start mewing. This cacophony can last from May to July. To humans we all look the same but my father was much larger than my mother. We gulls have little social contact with other gulls even with our relations and children. Our relations may extend to over ten generations which live close by or even next to us.

During his initial courtship my father would pretend he was a young bird and lower his head and give juvenile begging calls. I don't know if that impressed my mother or not. He then would give a supportive mewing call turning his head and if he had food would regurgitate it and give it to her. With all this food deliverance he would lose weight. If my mother accepted it then she became his for life and ready to lay eggs at three years of age. There were lots of gulls returning this year looking for mates so the noise was especially loud. The sky was full of gulls. Single gulls often give mewing calls between bouts of trumpeting.

My mother would always choose her nest site in May and she would return there, as would her mate, every year. My father was mainly involved in building our nest. It was not very deep and even included bits of plastic and string. He then added more vegetation and grass cuttings from the garden after the eggs were laid.

My mother would always return before the younger birds to our nesting site. Over the road one could see a male frantically trying to build nests on two houses. On each house was a female but no other male was in sight. Both females may lay eggs but with this work stretched male one set may not have been successful in hatching or chicks surviving. Certainly, taking turns in sitting on eggs would pose a problem. If there are too many females then their females may even pair up together.

My mother laid three eggs over six days. After egg laying my mother could be fed with marine invertebrates and fish by my father. She incubated our eggs for a longer time than my father during the night. I was one of three eggs laid behind a chimney stack on a housing estate. If we lost one then my mother would have to lay another. I hatched after three weeks as a small grey fluffy ball in June and after a week was running around. My brothers hatched at four weeks. My father had thrown the broken egg shells out of the nest in case other predators were attracted by the bright white inside. My parents were always arguing but very caring and very defensive ensuring I stayed in the nest. I would venture outside the nest and would often find myself slipping down the sloping roof only to be chased back into the nest by my parents.

During July, I tried learning to fly by flapping my wings up and down but they were so small and covered in fluffy feathers so I could get no lift at all. When one of my parents returned with food we would fight over it, all trying to tap our parents' beaks at the red spot to encourage them to disgorge their food. This was difficult and I often went without food for extended periods of time. Well, it seemed long to me. We were fed by our parents for at least eleven to twelve weeks both night and day. Often my last brother to hatch would get less food as we had developed sooner and possibly were a bit more aggressive. Eventually our meals became shortened and we searched for our own food. As a chick, my father provided most of the food and then it was up to my mother to feed me as a fledgling. We were taught things by our parents and could see how they reacted to different situations. I was born here and am destined to return to this area year after year.

We edged our way down the roof, all three of us. Our parents this time did not attempt to chase us back up to the nest. We learnt how to fly a bit by continually slipping down the roof and flapping our wings to get us back up. Our parents circled around above us calling and my father landed on the grass in front of the garden. For some reason, possibly my flat feet, while walking along the gutter I fell as did one of my brothers. I fell flat on my side but I picked myself up and wandered over to my father and begged for food. I could see that some of the other chicks from the other houses had not been so lucky and their remains had been embedded into the tarmac by a passing car.

As seagulls, we seem not to have evolved the sense to avoid cars like many of the other birds. We even stand on top of car roofs in car parks. Many of us fledglings are destined to die in this urban environment.

By the end of July we were all ready to follow our parents down to the beach or other open grassy areas. You would recognize many of us by the brown streaks on the head and neck. Unfortunately we wouldn't get our lovely adult plumage until we are four years old. Many of us young ones are also found in areas close to restaurants near the sea so that our parents might be able to get discarded food to feed us.

 If no chicks are produced then the parents may split up and look for other partners. It is not easy to find other mates and they may be rejected because of their previous divorce.

As a chick, I made high pitched squeaks and would raise and lower my head in a submissive posture. This worked sometimes but my parents could ignore me. I would even do it in front of humans hoping for some scraps and this worked too. If it got too hot my parents would show me how to pant as I had no sweat glands.

In the past there were cyclical rhythms of fish shoals linked to the seasons arriving near the shore. Their presence attracted thousands of fishermen. Even from Roman times this seasonal abundance of fish was recognized and became an essential resource for many people especially in Europe. Herrings are an example of a huge abundance of fish and formed an important food for Catholics on fasting days when meat could not be eaten. Miles and miles of herring swam in huge shoals. They were very popular in Victorian time when they were smoked. Fish processing factories provided us with food especially when disposing of offal.

In the twentieth century tens of thousands of herring girls in teams of three spent the summer and autumn in all weathers removing the guts and packing herring into barrels. They started in Scotland and travelled east following the arrival of the huge shoals. We were able to feed well then. Girls were situated either side of the raised troughs filled with herring which were sprinkled with salt. Fish guts were quickly removed by turning the wrist with a sharp knife. Guts were thrown into a gut basket and the herrings were sorted by quality and age. Fish were graded by size and if they had eggs. Those having released their roe were thin fish found at the end of the season. Tombellies were damaged fish, often thrown away for us to feed on. Herrings in the barrels were packed by the tallest girl and covered in salt then sent for export.

Herrings were sold in barrows on the street and transported on herring trains all over the country. When fishing, herring were trapped by their gills in the miles of netting and many fell into the sea to be picked up by us gulls. Thousands of fishing boats fought to get into the harbours with their catches. There were plenty of cast offs for us seagulls to eat, especially where herrings, cod, whiting and sprats had been filleted on ship and tons of offal discarded into the sea or unloaded on shore.

The Second World War finally brought about the end of the herring industry and the herring girls were no longer necessary. And our food resources seriously diminished.

We are not fussy and can switch to different foods. Sea creatures as well as fish have been a valuable source of food for us. Sometimes there is a glut of some species washed into shore, such as starfish, which make a good meal.

Many of us can migrate to areas with good food resources at different times of the year. We can only dive down a few metres to get food from the sea but we often prefer crustaceans and echinoderms on the shore. Shell fish we can break open or drop them on the rocks. We follow whale fishing boats and look for squid on the top of the water.

Falling fish stocks have forced us inland where we still use our swooping behaviour to get food much to the annoyance of many humans.

Windows of opportunity have been created by

humans in their throwaway society. We have good night vision and can see ultraviolet rays but we do not always need this as street lamps give us the chance to look for food at night when the day trippers have left.

You humans have invited us into the city where we have developed a fondness for junk food. We have learned to take hold of the outside paper of discarded fish and chips and tear it open or shake it to get the chips out. We have been trained to look out for food, the dropping of litter, or other indicators of food. Often food is wasted at parties even feeding on the guests' vomit. We can recognize the sound of lorries and get visual clues which indicate a human throwing an object or fish splashing in the sea. We have good eyesight and flying over an area we can see things that look like food.

Humans feed us in the garden and elsewhere. Some people regularly go down on the beach every evening to feed us. We can form special relationships with people. Children like the ritual of feeding us. We have begun to expect humans to share their food with us. We have learnt to tap on windows and glass doors to ask for food. We hang around food vans where left overs are left on tables or thrown on the ground.

People buy food from supermarkets to feed us and know what we like such as cat biscuits. We can arrive in hoards and can arrive every day at these times, especially at landfill, and squabble over the food. We need to arrive early but may lose out by calling the others.

We have learnt how to peck open sauce packets used in restaurants. We can easily tear open black plastic sacks, often spreading the rubbish across the street or road and the food remains are gladly eaten.

When desperate we paddle like Irish step dancers on the beach as the tide recedes with our rain dance which helps to bring up shellfish. On grass we use the same method to bring up worms. We search on the beach for crabs in pools and steal other birds' eggs. It has been known that if we have lost our own brood we may eat eggs or even chicks from other sea gull nests. We have been known to feed on carrion such as rats and dead rabbits or even fruit falls. We sometimes eat roots or seeds. But we really prefer animal foods Things we cannot digest we regurgitate. We stay all day at refuse tips and follow the tractors ploughing up fields to get the worms and other exposed grubs.

We drink fresh water out of ponds or lakes, birdbaths or water left in roof gutters or recesses. We can drink sea water but we lose some of this salty water from our nostrils so as not to overload our bodies with salt.

Infections picked up by surface dipping across sewage outflows eating fish offal or other waste together with ingesting plastics, has caused a decline in our numbers. Old members of our group often die of food shortages as well as chicks that cannot feed themselves.

5 ARE WE REALLY A PEST OR A THREAT?

A lot of birds are considered pests in towns and cities especially when in large numbers. At one time there were lots of sparrows in Leicester Square and pigeons in Trafalgar Square and St Mark's Square in Rome but these are few and far between these days.

Marabou storks live on high buildings in Africa and India and have no predators. They are scavengers seen feeding around rubbish heaps. These ugly birds were at one time an object of revulsion especially when seen feeding on cadavers during times of civil war. They were believed to bring bad luck and were hunted and killed but now are revered by some cultures and are targets for conservation. Some people have customs such as feeding birds during funerals as they believe we are linked to spirits of the dead

Our reputation and attempts at removal is often linked with stealing food while people are eating or swooping down when we see a threat to our fledglings. On the other hand they make a big fuss if we are injured or entangled in fishing line with hooks in our mouths. We can't avoid dropping faeces onto your cars or buildings and also causing a commotion at 4.00 am in the morning. We have little effect on polluting the water, much less than the dog poo left on the beach in the winter remaining hidden in the sand.

With recycling and dwindling waste we have looked to man again for other windows of opportunity for us. In order to survive some of us expect humans to provide us with food. We see food not the human holding it. We know we can gain a lot more food using this method than by just begging. Our sense of sight is very good as is our interpretation of sound but our sense of smell is not great. We reckon women more are likely to drop food if attacked. It is essential for us to find food when our young are born and for several weeks after when they still can't feed themselves so we will do anything to get food. We have even picked up a man's false teeth when he took them out to eat a biscuit. Once we have found a food source this can encourage other interested groups to join us. Weekends and nights are popular for scraps from takeaways. We beat the street cleaners in the mornings and find lots of scraps often hanging out of full bins. Wave even learnt that we can get food after the school bell has rung for lunch. We have real problems where there are exorbitant fines for feeding us.

We also have to protect our young during our nesting time, April to October. People need to be patient with us, it's just our nature. We know from the past that there are dangers when animals and humans are around at peak holiday times, especially for our young when they have just flown down from their nest. Just as any parent would be concerned if they saw a threat to their children.

We can dive down like a dive bomber at 40 km/hour hovering above to ward off intruders. We are not the only birds to do this. Terns and others can mob intruders near their nesting sites. We can even spray our faeces or vomit over them. This usually shoos them away but persistence especially a barking dog, can initiate an attack by one or other recruited gulls and may even result in serious injuries caused by our talons and beak of some our more aggressive relatives. We can make ourselves bigger by withdrawing our necks into our body which may frighten some animals away.

To remove us or stop us breeding they have used a whole range of methods. They can destroy our eggs by piercing the shells or coat our eggs with liquid paraffin so that the developing chicks can't breathe. On buildings they put up plastic owls and owl shaped kites but we get used to them. They also broadcast loud irritating noise or even use firecrackers which bang at intervals to scare us. They use spikes, wires, nets, lights, rotating streamers, coloured bunting, electric zappers, and water pistols for people to use when eating outside. They even think that the colour red can deter us.

Also they release peregrine falcons to attack our young in our nests or fly around football stadiums before the matches.

To kill us they have shot us. The use of poison they feel may kill other animals so it is voided, The 1981 Wildlife Countryside Act says it's illegal to capture, injure, destroy or interfere with our nests and eggs but people can get a general licence if they can prove we are a danger to public health or safety.

If threatened with an attack or dive bombing all people need to realise is that we are attacking irrespective of who they are, man or animal. Raising their hand or lifting something above their heads can help reduce any mobbing. We often go for the tallest person present or the highest point, hence the raised hand. We don't want to hit you and will fly off. If you eat food next to a wall or building or under an awning we cannot attack you. Once you experience us swooping down you will be more wary next time.

We will be around for a long time yet, a mascot for the seaside, noisy and raucous as usual, hungry and desperate for food, especially with our young to look after. You have fed us for many years and please continue to do so as well as feeding other birds.

To see other publications below by the author visit
snappysnappybooks.com

<u>The really, really, really useful series</u>

How to be a Successful Business Weed
How to Deal with Life's Snakes and Ladders
Know Your Students and Build Your Image
Pens for Pops
How to be a Successful Charity Shop
Make up-revealed
Ronnie's Sermon snippets
Wastefulness-Bone and Urine
Fertility Stones and Chocolate Eggs

<u>Other books by Mike Pearce:</u>

Pattern for Purpose- God's and Man's designs
Red Fred Cell and Friends
Human Termites eat London
Pigeons Splat London
Glass Anemones Tentacle-ize London
Tuppeny Hangover
I am Termite
The littlest Oyster
Bits and Bobs
The Shell Man
Cats at Christmas
Tails, Tales
Trust-Nothing but a Must
In a Dark, Dark Corner was the Holy Ghost
The Shell Lady
Captain Grottbuster versus the Grey World
London's Nemesis (Trilogy of 3, 4 and 5 above

Saved by Angels (Trilogy of 6, 8 and 14 above)
The World of Wax
Photosynthetic Women
Queen Rat on Deadman's Island
The Watcher on the Fal
The Rock Pool
The Little Shepherd Boy's Gift
The Living Fossils
Old Mother Nature Laughed and Laughed
Betty's Barcodes
Time Runs Dry (play)
Valentines Cards
The Scrofula Infirmary
The Cornish Urchin
My Therizinosaurus
Spider in the Tomb
The White Cockerel
The Red Church Doll
Butterfly Angels (compilation of previous books)
The Girl Under the Paeony Tree
Baby Feet
The Sparrows' Last Soul

ABOUT THE AUTHOR

Dr Mike Pearce is a scientist interested in behaviour.
He also was a lecturer in human biology and health at
a college in Canterbury, Kent.

Printed in Great Britain
by Amazon

46119748R00031